On Action and Reaction

Greg Feild

September 24, 2017

The final test of deduction lies in experimental observation.

Elaboration by reasoning may make a suggested idea very rich and very plausible,
but it will not settle the validity of that idea.

Only if facts can be observed (by methods of collection or experimentation),
that agree in detail and without exception with the deduced results,
are we justified in accepting the deduction as giving a valid conclusion.

Thinking, in short, must end as well as begin in the domain of concrete observations,
if it is to be complete thinking.

-- John Dewey
How We Think

Abstract:

In this penultimate paper on the universal model of our world,
we take stock, and examine the current status of our model in
the spirit of our two previous efforts; "On Spin, Mass, and Charge"
and "The Sinister Universe".

We will refrain from (*too*) much new speculation,
so as to not muddy the waters.

However, we will investigate several of the unresolved issues and new
proposals highlighted in our last paper, "On Matter, Mass, and Motion".

About the author:

Greg Feild is a 'gentleman' scientist
and an armchair philosopher.

His goal is to make physics fun again ...
and physical!

Coming soon:

"A Quantum Mechanical Theory of Everything" !

:)

Preface:

Our new universal model of the world actually consists of many separate, and seemingly disparate new theories, that are however, logically intertwined; both suggesting and reinforcing one another. It basically explains every outstanding problem of the standard model and more! A bold statement to be sure.

Having started our enquiries 'from scratch', with a broad view of the successes and perplexities of all current models, we were able to confront and dispose of these problems, taking one at a time, with 'relative' ease.

Bold as love!   :)

Now, whether our new model is asymptotically approaching coherence and/or completeness, is another question, and one, ultimately, not to be decided by ourselves!

In the brief synopsis that follows, we assume the reader is a fan (or frenemy) of our new model, or is at least familiar with our previous 12 books.

The 'universal model' can be broadly characterized, or divided into, five complementary (and traditionally separate or distinct) 'sub-theories';

1) A quantum and cosmological theory of gravitational interactions
2) A classical and quantum mechanical theory of all particle interactions
3) A theory of time and space
4) A theory of the fundamental nature of elementary particles
5) A physical interpretation of the quantum mechanical wave function

In our model, gravitation appears to be a 'carbon copy' of electromagnetism, except for the lack of a negative gravitational charge.

Otherwise, gravity and electromagnetism differ only by a factor of *scale(s)*
(i.e. by factors of G, $e/4\pi\varepsilon$, and $1/c^2$),

The fundamental coupling charge, for both quantum and cosmological interactions, in both gravitational interactions *and* electromagnetic interactions, is the *total* relativistic mass of the particle or planet involved.

From this realization alone, most of the theory falls almost magically into place.

We now present a 'laundry list' of the issues we believe are explained by the model:

- Matter-antimatter asymmetry
- The nature of antimatter
- Matter-antimatter annihilation
- The origin of parity
- Neutrino handedness
- The nature of the strong force
- Why the proton charge is exactly e
- Parton confinement and asymptotic freedom
- The weak force
- The origin of mass and charge
- Muon decay
- Particle families
- Gravity
- The $1/R^2$ law
- Wave particle duality
- The wave function
- The nature of the photon
- The nature of leptons
- The spinor nature of the electron
- SU(3)
- Gauge theory
- The running of alpha
- Beta Decay
- Why the Born approximation is exact
- Issues in cosmology
- Relativistic mass
- Dark matter
- Dark energy
- Quantization
- And much, much, more !

Introduction:

In our first paperbook "A Quantum Mechanical Theory of Gravitational Interactions", we began our enquiries by proposing a modified principle of equivalence;

    inertial mass = gravitational mass = gravitational charge   (1)

Our recent investigations suggest an even *more* general expression for this principle, which we shall call the 'universal principle of equivalence'.

    relativistic inertial mass = relativistic gravitational mass = gravitational charge

                  == *the fundamental universal coupling charge*   (2)

In our new model, the "relativistic" particle mass is the gravitational charge, the "strong force" charge, the weak charge, and ultimately, the electric charge as well!

The universal theory of relativity:

Why do we use "scare quotes" when we say the 'relativistic' mass?

    The postulates of the special theory of relativity are;

        i) Physical laws are the same in all inertial reference frames.
        ii) The speed of light is a universal constant in any reference frame.

As an alternative, we propose the postulates of 'the *universal* theory of relativity';

        i) Particle interactions occur at the speed of light in any reference frame.
        ii) Particle interactions obey the $1/R^2$ law in any reference frame.

        iii) $F = d\mathbf{p}/dt$

In an inertial reference frame, I believe, these three postulates should lead one (the challenged reader!) to the familiar expressions for the relativistic mass and momentum, *without* reference to, or resorting to, coordinate transformations.

The familiar Lorentz transformations would then follow naturally as a consequence.

The speed of light:

All particle interactions occur at the speed of light. Photons are interacting particles that travel at the speed of light. Photons have the same speed in all inertial reference frames.

However, photons do not have the same energy, or frequency, in all inertial reference frames.

This last point is 'obvious', but it also seems very important somehow ...

We will use this observation to establish a universal reference frame, absolutely at rest; a fixed background that is always 'at rest' relative to any and all motion.

The universal reference frame:

An inertial observer is *defined* to be absolutely at rest, against the fixed background of space, if they measure the accepted, absolute value for a particular and well known wavelength of the cosmic background radiation.

An observer in a *moving* inertial reference frame, could then determine their absolute velocity relative to the fixed background of space, by measuring the Doppler shift of the cosmic microwave background.

The cosmic microwave background:

So, what is the cosmic microwave background?

We've essentially ruled out "stretched" photons due to either, a) the big bang, or b) an eternally expanding and contracting universe.

Our current model suggests a static and eternal universe.

So, the cosmic microwave background must be gravitational bremsstrahlung!

A collection of soft photons, emitted by every massive particle, ever accelerated, ever!

The Lorentz force:

In "On Matter, Mass, and Motion', we found it convenient to express the Lorentz force between two bodies as a function of the total energy of the system;

$$F/E_{TOT} = K*(c/R)^2 \mu (r + (1/c^2)(v_1 x v_2 x r)) \qquad (3)$$

where $\mu(R, dR/dt) = m_1 m_2/(m_1 + m_2)$ is the reduced mass of the two body system, and $r = R/R$. The reduced mass, is the relativistic mass, and depends on the position *and* velocity of the particles. The constant K consolidates all the coupling constants;

$$K == (G/c^2 - (e/m_e)^2(\mu/4\pi)) \qquad (4)$$

We interpret the second term in equation (3) as a generalization of the familiar coriolis force which arises when studying the motion of bodies on earth from a fixed reference frame.

In our generalized Lorentz force, the coriolis force is not an artifact of the choice of a particular reference frame, but arises from the absolute relative motion of the two bodies.

In our model, the coriolis force is *real*, because all forces are velocity dependent due to the fact that particle mass is velocity dependent; $m = m(r,v)$.

Similarly, the centrifugal force is now also a real force and is not due to "space time" disturbances as in the general theory of relativity

Equation (3) should be 'easy' to generalize to an N body system.

The Lorentz torque:

The 'Lorentz torque' arises from the interaction of the intrinsic angular momentum (quantum spin or classical moment of inertia) of one body with the magnetic field vector of the second body, yielding the force F_{SPIN} introduced in "On Matter, Mass, and Motion".

We still aren't ready to work out the formula for F_{SPIN} (readers?), but we do note that the magnetic force *now does work*, and this force is equal, opposite, and *central* between the two bodies.

It seems the extra, work free, looping and spiraling, circular motion of a charged particle in an external magnetic field is one of Nature's red herrings! Artful window dressing.

The classical propagator:

In our last book, we challenged the reader (and ourselves!) to create a 'propagator' from our new universal Lorentz force.

The complete, 'classical', relativistic, Lorentz force between two identical electrons is

$$\mathbf{F} = (G/c^2 - (e/m_e)^2(\mu/4\pi))(c/R)^2(m_1 m_2 \mathbf{r} + (1/c^2)(\mathbf{p_1 x p_2 x r}))$$

where, of course, $\mathbf{F_1} = -\mathbf{F_2}$. We shall start our study slowly, and consider only the static, or Coulomb potential, as one usually does.

$$\mathbf{F}_{1,2} = K*(c^2/(\mathbf{r_1} - \mathbf{r_2})^2 (m_1(\mathbf{r_1}, \mathbf{v_1}))(m_2(\mathbf{r_2}, \mathbf{v_2}))$$

The intriguing factor of c^2/R^2 (the result of some factoring), already looks 'propagator like', with units of $1/t^2$.

We could look at the work

$$W_{1,2} = -W_{2,1} \Rightarrow W_{1,2} + W_{2,1} = 0$$

or even the action. I believe the key to this problem is converting the integrals such that we are integrating over the masses, dm, of the two particles; the limits of integration being the initial and final relativistic masses of the two particles.

We even think we have a proof, but unfortunately it will not fit in this giant space below.

The electron:

In "On Matter, Mass, and Motion", we derived/induced the universal model generalization of the formula for the magnetic moment of the electron;

$$\mu_e = (e/m_e)(m)(\hbar/2m_e) \tag{5}$$

That is

$$\mu_e = (\text{coupling constant})*(\text{mass})*(\text{angular momentum per unit mass})$$

So, the electron is one unit of angular momentum per unit mass per unit volume of space; or one unit of inertial, half integral, angular momentum per unit mass.

$$e \Rightarrow h/4\pi m_e == (\hbar/2)/m_e == L/m_e \tag{6}$$

Inserting the formula for the relativistic mass into equation (5) we get

$$\mu_e = (e*\hbar/2m_e)(1/(1 - v^2/c^2)^{1/2}) \tag{7}$$

We then make the usual series expansion to obtain

$$\mu_e = (e*\hbar/2m_e)(1 + \tfrac{1}{2} v^2/c^2 + \tfrac{3}{8} v^4/c^4 + ...) \tag{8}$$

and find the magnetic moment of the electron increases with velocity, *as expected*.

The electron is essentially the vector **L**/m_e. This vector precesses about the axis of the direction of motion with a frequency; $\nu = E/h = m/h$. As the speed of the electron increases, the frequency of precession increases, and the mass of the electron increases.

The surprising 'spinor' nature of the electron is due to the angular momentum vector flipping helicity/polarization every 2π radians.

Even though the mass and magnetic moment of the electron increase with the velocity, the electron angular momentum is always $\hbar/2$.

We finally know what has been waving all this time!

The waving of the wave equation/wave function represents the periodic precession of the electron spin about the direction of travel of the electron!

N.B. The wave equation describes *many* physical phenomena that don't really wave.

Spinoring:

We've been saying the "helicity or polarization" of the electron 'flips' every 2π radians.

Technically, of course, this terminology is incorrect, because even though there *is* 'flipping' going on, the helicity of the electron never changes!

Instead, we shall choose to say the electron is 'spinor-ing', as illustrated in Figure 1.

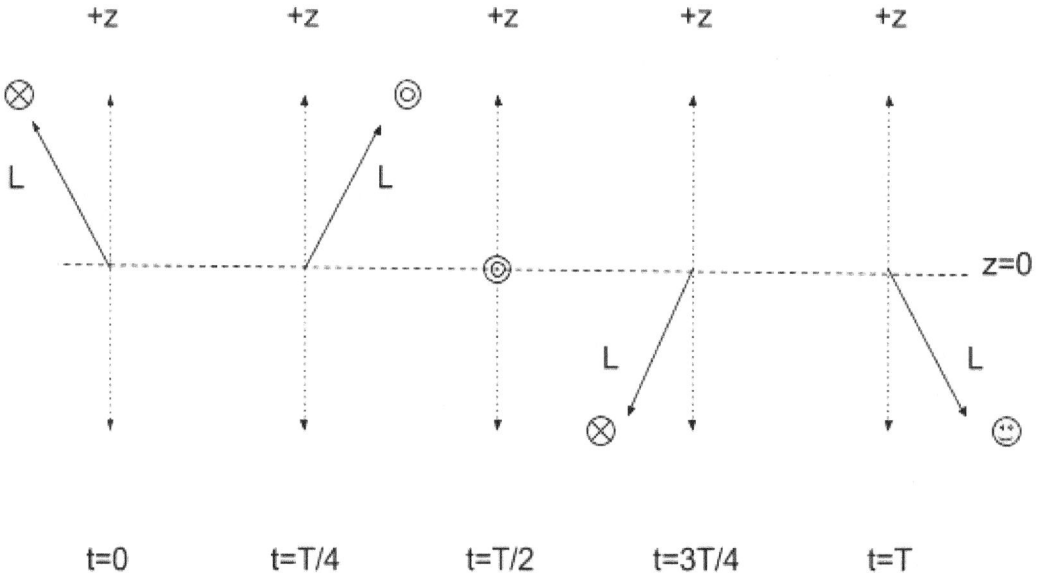

Figure 1: At rest with an electron traveling the the z-direction. The spin angular momentum vector 'precesses' about the direction of motion, tracing out a closed, three dimensional figure eight (a string!). The x symbol represents motion into the page. The dot symbol represents motion out of the page. At time T/2, we see the the angular momentum is *perpendicular* to the direction of travel. (This is when the electron engages in 'virtual' interactions.)

We also can see, that although the angular momentum vector is 'spinor-ing', the polarization, or helicity, of the electron does *not* change, and is constant!

Whew!

The neutrino:

We believe (really, really want) the electron neutrino to have a magnetic moment. Actually, we are (almost) *sure* it does, and we are determined to figure out!

This hypothesis is a core premise of our model! So, consider the electron ...

Even though the electron is electrically charged, we can see from equations (5) and (8), that the magnetic moment is really only dependent on the mass of the electron.

The electric charge, e, functions solely as a proportionally factor or *coupling constant*.

Of course, this has been our premise since "On Gravitation and Electric Charge".

Now, we will construct the neutrino magnetic moment in strict analogy with equation (5)

$$\mu_\nu = (m)(\hbar/2m_\nu) \qquad (9)$$

Originally, way back in "Observations on the Quantum Mechanical Nature of Gravity", we expected or assumed the neutrino magnetic moment would have the same units, or "charge", as the electron magnetic moment, involving the electric charge, e.

In hindsight, we can see this approach was wrong for two reasons

1) The gravitational magnetic vector term, **B_g**, of our new Lorentz force does not depend on the electric charge.
2) We have already checked that the 'cross term' in our expression for alpha_strong, as derived in "The Sinister Universe", is dimensionally correct.

If we take equation (9) as the magnetic moment of the neutrino and form the vector dot product with either **B** or **B_g** from our generalized Lorentz force, we see the units also work out correctly.

So, we have *finally* figured out the magnetic moment of the neutrino!

Whew.

(We may have just achieved complete coherence!)

Inserting the formula for the relativistic mass of the neutrino into equation (9) we see

$$mu\_v = (hbar/2)(1 + \tfrac{1}{2} v^2/c^2 + \tfrac{3}{8} v^4 c^4 + ...) \tag{10}$$

And, we must conclude that *the neutrino is one inertial quantum of action* as we proposed in "On Matter, Mass, and Motion". So, we can represent the neutrino symbolically, as we did for the electron in equation (6), and write

$$nu\_e \Rightarrow hbar/2 == L \tag{11}$$

Actually, equation (10) shows that *there is only one neutrino*. The three neutrinos differ only by their **velocity.**

The tau lepton is the most massive of the three charged leptons, so when it 'decays', it emits the highest velocity, and hence, most massive of the the "three" neutrinos, usw.

The tau is shedding energy and spin, because what else is there?

We can also derive the magnetic moment of the neutrino using the hand waving arguments usually assumed for the electron magnetic moment.

$$mu = (mass\ current)(area) = (m/t)(A) = m(L/2m) = hbar/2 \tag{12}$$

Finally, we reach way back to Physics 101 and recall the formula for angular momentum

$$L = mvr \tag{13}$$

The angular momentum of the neutrino is h and we assume it spins with angular velocity c.

$$h = (m\_nu)(r\_nu)c \tag{14}$$

Now, we can solve for the *Compton radius of the neutrino*; the smallest probable distance.

$$r\_nu = h/(m\_nu)*c \tag{15}$$

The photon:

The photon is essentially a perpetual motion machine! The photon polarization oscillates harmonically at a frequency proportional to its energy, nu = E/h, as discussed in "On Matter, Mass and Motion". This is an unorthodox picture of the photon polarization, but we note it satisfies the orthogonality condition on the photon wave vectors; $\mathbf{k} \cdot \varepsilon = 0$.

The photon is one 'free', massive, but inertialess, unit of angular momentum; L = h.

The neutrino is one 'bound', massive, unit of angular momentum *per unit space*; L = hbar/2.

The electron is one 'bound' unit of angular momentum per unit space *per unit mass*; L = hbar/2.

Particle interactions:

Take a look at the angular momentum vector of the electron at time, t=0, in Figure 1. Here, we say the electron helicity is in 'full bloom' and it is able to absorb a real photon (*if* its polarization is also 'blooming'), increasing the rate of precession of the electron angular momentum vector, and hence the electron mass.

Consider the case of partial transmission and partial reflection of light from a thin sheet of glass. Photons in 'full bloom' will be transmitted. Photons with polarization ~0 will be reflected.

Similar arguments can be made for electron tunneling. If an electron arrives at a potential barrier 'out of phase for reflection', *and,* the electron 'wavelength' is comparable to the 'height' of the potential barrier, the electron will 'tunnel' through!

Quantum field theory:

In our model, particles are not created and destroyed. Instead, particles absorb, emit, merge with, and *shed* one another.

For example, an electron and a positron do not 'annihilate', producing a virtual photon.

The electron and positron have equal and opposite spins, ½. They merge to become a photon of spin 0. The photon then splits into the particle antiparticle pair demanded by the situation.

In our model, particle interactions are a continuous flux and flow of energy and momentum, *flowing only one way*; futureward. Particles interact by exchanging units of *angular momentum*.

13

Conclusion:

I apologize for all the homework!

However, building a model is a collaborative effort, as we've emphasized before.

:)

And, math is hard! It would take me a very long, frustrating while to (try and) work through all the math we have suggested. So, I leave it to the professionals!

I would much rather get the core ideas out to the community, than try to be a 'hero' and work through all the math myself.

There may be still some 'issues' with the model, of course, but only one person has been working on it, and for only one year! We need more people power and hours.

I believe this model, and it's several theories, is essentially correct.

I *know* there are plenty of good bits.   (Maybe even some 'clickbait' !)

Unfortunately, I am an 'outsider' (even though I have a PhD in physics, twenty years experience in the field, am the author of many conference proceedings, and the principal author of at least 5 physics papers published in peer reviewed journals, and the author of literally thousands of papers as the member of two international experimental collaborations).

But, rant over.   :)

Someone will discover my little books.

Someday.

They will be *my* hero!

Outsider physics!

home grown

References:

Modern Elementary Particle Physics
Gordon Kane

Classical Dynamics of Particles and Systems
Jerry B. Marion

Foundations of Electromagnetic Theory
John R. Reitz, Frederick J. Milford, Robert W. Christy

Quantum Physics
Rolf G. Winter

Gauge Theories in Particle Physics
I. J. R. Aitchison and A. J. G. Hey

Quarks and Leptons: An Introductory Course in Modern Particle Physics
Francis Halzen, Alan D. Martin

Quantum Field Theory
F. Mandl, G. Shaw

Theoretical Mechanics of Particles and Continua
Alexander L. Fetter, John Dirk Walecka

and

Elementary Modern Physics   (Best Book Ever!)
Richard T. Weidner, Robert L. Sells

a feild theory   :)

Books by Greg Feild:

1. "A quantum mechanical theory of gravitational interactions"
   CreateSpace Independent Publishing, 8/29/2016

2. "Observations on the quantum mechanical nature of gravity"
   CreateSpace Independent Publishing, 10/8/2016

3. "On gravitation and electric charge"
   CreateSpace Independent Publishing, 10/29/2016

4. "On spin, mass, and charge"
   CreateSpace Independent Publishing, 11/29/2016

5. "On angular momentum, acceleration, and absolute motion"
   CreateSpace Independent Publishing, 1/1/2017

6. "The Sinister Universe"
   CreateSpace Independent Publishing, 3/1/2017

7. "On Parity and Isospin"
   CreateSpace Independent Publishing, 4/11/2017

8. "Reflections on the Sinister Universe"
   CreateSpace Independent Publishing, 5/12/2017

9. "On Current Physics"
   CreateSpace Independent Publishing, 6/11/2017

10. "A Critical Examination of Classical and Quantum Mechanical Waves"
    CreateSpace Independent Publishing, 6/18/2017

11. "On wave particle duality and the quantum of action"
    CreateSpace Independent Publishing, 7/6/2017

12. "On Matter, Mass, and Motion"
    CreateSpace Independent Publishing, 9/14/2017

Compilation:

"The Universal Model of Our Sinister Universe: The First Ten Books"
CreateSpace Independent Publishing, 7/2/2017

a greg feild production

gf

Notes:

There was an old lady who ate a fly:

There was an old lady who ate a horse.

She died, of course!

an allegorical tale

a parable for our times!

keep it simple!

physics is fun!!

:)

www.ingramcontent.com/pod-product-compliance
Lightning Source LLC
Chambersburg PA
CBHW082225220526
45470CB00010B/3315